I0533295

© 2025 **Boricua Science**

International Standard Book Number (ISBN): 979-8-218-68947-6
Library of Congress Control Number (LCCN): 2025910219

First edition, 2025
Printed in the United States of America

Cover and interior illustrations co-illustrated by Dr. Aidyl Gonzalez-Serricchio
Design and layout by Dr. Aidyl Gonzalez-Serricchio (Google Docs) and Joey Serricchio (Adobe InDesign)

For information or permissions, contact:
Boricua Science
www.draidylgonzalez.com
aidylgonzalezwebsite@gmail.com

Everyday Lab Math™ and "*Science moves.*™ *So do I.* are trademarks of Dr. Aidyl Gonzalez-Serricchio.

Author's Note / "Why I Wrote This Book"

Hola, My Fellow World Changers,

   I used to think math wasn't for me.

Growing up, I struggled with numbers, especially the abstract kind. As someone who learns differently, math often feels like a closed door. But everything changed when I stepped into a lab. That's when I realized something big: **math with units isn't abstract**. It's real, it's tangible, and it powers everything we do in science, from mixing buffers to measuring DNA concentrations to creating that yummy dish.

Lab math is not about memorizing formulas. It's about making sense of the world, solving problems, and doing accurate and repeatable science.

This book is for anyone who felt math was out of reach. **You belong here** whether you're a high school student, a biology major, or just someone curious about how science works behind the scenes. This is your guide to understanding the essential math used every day by biologists and chemists in labs around the world.

Follow along, take your time, and trust that you can do this because you can!

Take Care, Be Well, and Stay Awesome,
Aidyl

How to Read This Book (Even If You Hated Math Class)
Your roadmap to mixing buffers, serums, and science with confidence

This book is built for how real people learn through doing, visualizing, and making sense of science in everyday life. To make your journey even smoother, I've added:

1. Chemistry Flow Maps (*"Where Am I and Why Does This Matter?"*)

Before you dive into Chapter 2, you'll find a **mini-map** connecting the key acid-base concepts: **Acids, bases, pH, water ionization, and chemical equilibrium** aren't just topics; they're parts of one giant molecular soap opera. The visual map gives you a preview of how they interact, so even if you're skimming for a specific formula or troubleshooting a recipe, you'll know where you are in the story.

2. Natural Recaps Moments

Rather than boxed summaries, this book includes built-in pause points moments where we zoom out, connect dots, and reflect on key ideas. They're embedded in the flow with clear cues like "Why This Matters," "DIY Insight," or "Take-Home Message," so you'll know when we're stepping back to synthesize before moving forward.

3. Resource Pages and Bonus Downloads

In the back, there's a handy **Quick Reference Guide** with:

- Common lab math formulas.
- Units and conversions.
- pH, Ka, and buffer equations.
- A cheat sheet for % solutions, dilutions, and more.

Bonus: Use the *QR code* to download a *printable version* of the reference chart, perfect for taping to your lab notebook or skincare cabinet.

Table of Contents

Chapter 1: Solution Avengers Assemble – Molarity, Percentages, & the Power of the Mole

Science doesn't happen without math. And in the lab, the kind of math that matters most? The math of solutions. Don't worry this chapter breaks it all down.

Metric System Quick Refresher

Prefix	Symbol	Meaning	Example
kilo	k	1,000	1 kg = 1,000g
centi	c	0.01 or 10^{-2}	1 cL = 0.01L
milli	m	0.001 or 10^{-3}	1mL = 0.001L
micro	μ	0.000001 or 10^{-6}	1 μg = 0.000001g
nano	n	0.000000001 or 10^{-9}	1 nL = 0.000000001L

A. What is an Aqueous Solution?

An aqueous solution is a **homogeneous mixture** where a solute is dissolved in water.

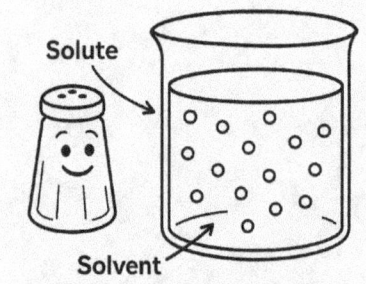

AN AQUEOUS SOLUTION

Solute

Solvent

A homogeneous mixture where a solute is dissolved in water.

Solute: the substance being dissolved (e.g. salt)

Solvent: the substance doing the dissolving

- **Solute**: the substance being dissolved (e.g., salt).

- **Solvent**: the substance doing the dissolving (usually water).

B. Concentrations based on volume

1. Meet the Mole: Why 6.022×10^{23} Isn't a Random Number

A 1M solution contains **one mole** of solute particles which equals: **602,200,000,000,000,000,000,000 particles** or **6.022×10^{23} particles.**

This is known as **Avogadro's number**, and it applies to atoms, molecules, ions, and any other countable entity in chemistry.

So:

- 1 gram-mole (g-mol) = 1 mole = 6.022×10^{23} particles
- 1 gram-ion (for ions like K^+) = 1 mole of ions
- 1 gram-atom (for elements like Na) = 1 mole of atoms

Basically, one Avogadro's number of any particles, whether atoms, ions, or molecules, are called a "**mole**".

Think of it like this: when I say "dozen" you think of 12. When I say "mole" you should think 6.022×10^{23}.

2. Lab Math Starts with M (for Molarity!)

In the lab, concentrations based on the **amount of solute per unit volume** are used most often. Below are the most common ways this is measured and described. **Molarity (M)** = The number of moles of solute per liter of solution.

$$M = \frac{\text{moles of solute}}{\text{liters of solution}} \quad \text{or} \quad M = \frac{\text{grams of solute}}{\text{MW} \times \text{liters of solution}}$$

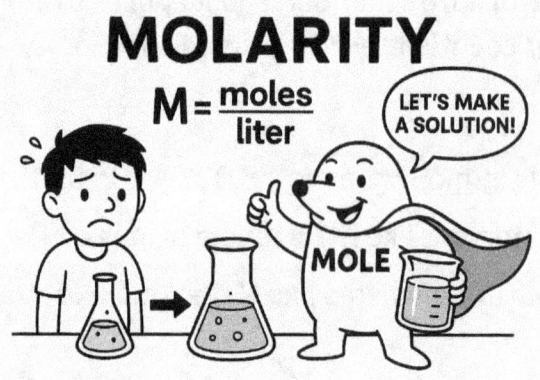

Molecular weight (MW) tells you how much 1 mole of something weighs in grams. You can usually find it on the periodic table or chemical label.

Molar concentrations are usually shown in **square brackets**. For example, [H^+] means "the molarity of hydrogen ions." Brackets are just shorthand chemists use to show "how much is in solution." To calculate **M** we need to know the weight of dissolved solute and its molecular weight, **MW**.

Where the subscript **g** refers to **grams**. Dilute solutions are often expressed in terms of millimolarity, micromolarity, and so one, where:

$$1 \text{ mmole} = 10^{-3} \text{ moles}$$
$$1 \text{ μmole} = 10^{-6} \text{ moles}$$
$$1 \text{ nmole} = 10^{-9} \text{ moles}$$
$$1 \text{ pmole} = 10^{-12} \text{ moles}$$

Dilute Solutions and Their Units

Small concentrations are often expressed in millimolar, micromolar, and beyond:

Unit	Conversion	Equivalent *
1 mM	10^{-3} M	1 μmole/ml
1 μM	10^{-6} M	1 nmole/ml
1 nM	10^{-9} M	1 pmole/ml
1 pM	10^{-12} M	1 fmole/ml

Equivalents assume a 1 mL volume. For example, 1 mM = 10^{-3} mol/L becomes 1 μmol/mL when scaled to milliliters.

Incredibly Important Fun Fact: Even a **1 μM** difference can affect the outcome of a reaction, so precision matters!

⌇ Citizen Science Spotlight: Soil Nutrient Test

Ever wonder what's really in your soil? You can find out using the molarity math you just learned.

Field Example: You collect 10 g of soil, mix it with 100 mL of distilled water, and filter the solution. After testing, you detect 0.002 moles of nitrate (NO_3^-).

What's the molarity? M = 0.002 mol ÷ 0.1 L = **0.02 M nitrate**

That's it! You just calculated real-world soil chemistry.

Want to do this as part of a global science project? Join our soil citizen science initiative and contribute your data to help scientists monitor soil health around the world. → **soilsciencelab.com**

3. Activity: Chemistry's Fine Print

Molarity tells you how many moles of solute are in a liter of solution but here's the catch: that number doesn't always reflect how the solute actually behaves in real life.

In a perfect world (*like the one your textbook sometimes pretends exists*), solutes dissolve, don't interact, and behave predictably. But **real solutions aren't always ideal.** Ions bump into each other. **Charges interfere. Concentrations get distorted.**

That's where **activity** comes in. It's a corrected version of concentration that accounts for these interactions, especially in solutions with high ionic strength. Think of it like a "*chemistry correction factor*" telling you how much of a solute is effectively doing the job you expect.

Activity is represented as: $a = \gamma \times [C]$

- a = activity
- γ = activity coefficient (ranges between 0 and 1)
- $[C]$ = molar concentration

When conditions are ideal, γ is close to 1 and activity ≈ *roughly equals* ≈ concentration. But in more concentrated or charged environments, the activity drops below the measured molarity.

You won't always need to calculate it in the lab but knowing that activity exists helps you understand why reactions might not follow the script, even when your math says they should.

Here's an exciting sample problem:

Hydrofluoric acid (HF) is a weak acid. In a **0.1M** solution, it doesn't fully dissociate. It behaves as if it contains only **0.039 M** of free H^+ ions. That means $\gamma = 0.039/0.1 = 0.39$.

Crazy, right? The activity is way lower than the molarity because the acid doesn't release all its protons. **MIND BLOWN!**

4. Normality: When One Mole Isn't Enough

Normality (N) measures the number of **equivalents of solute per liter of solution**.

$$N = \frac{\text{equivalents of solute}}{\text{liters of solution}}$$

What's an **equivalent**?

- For **acids and bases**, it's the number of moles of H^+ or OH^- donated or accepted.

- For **redox reactions**, it's the number of moles of electrons transferred.

   One equivalent is the amount of substance that can react with **1 mole** of H^+, OH^-, or electrons.

Here are useful examples:

1) H_2SO_4 (*sulfuric acid*) provides 2 ionizable H^+ ions per molecule. -> 1 M H_2SO_4 = 2 N

2) $Ca(OH)_2$ (*calcium hydroxide*) provides 2 OH^- per molecule. → 1 M $Ca(OH)_2$ = 2 N

How Normality and Molarity Are Related

Normality and *molarity* are *connected* through the number of reactive units per molecule, **n**. This formula makes it easy to convert **molarity** to **normality**, as long as you know how many H^+, OH^-, or electrons are involved.

$$N = n \times M$$

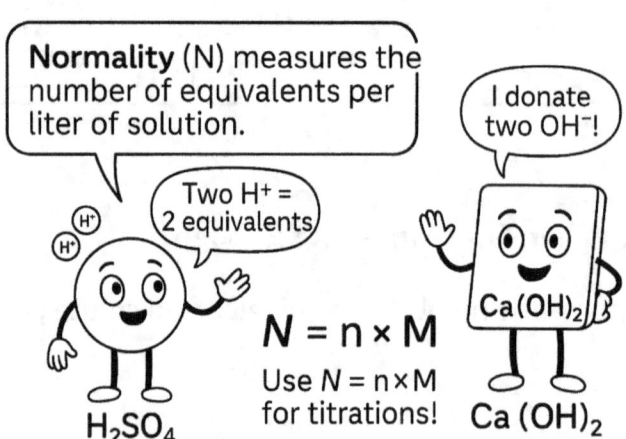

N= Normality; M= Molarity; n= number of equivalents per mole of solute

When is N important? **Normality** is especially helpful in **titrations** (i.e., those drop-by-drop showdowns where acids and bases face off until the chemistry is perfectly balanced) and reactions where charge balance or proton exchange is the key to success.

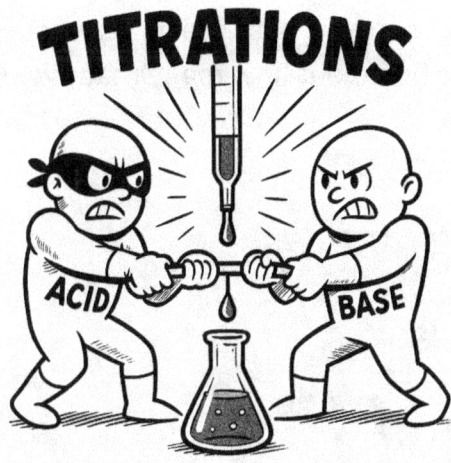

TITRATIONS

Lab Tip: When calculating **normality**, always check how many reactive units (H^+, OH^-, or e^-) the solute provides. (*Flip to Quick Reference Guide for the $N = n \times M$ shortcut*).

5. How Much Is Enough? Enter the Equivalent Weight

Equivalent weight (**EW**) is the **mass of a substance** that reacts with or supplies **1 mole of H^+, OH^-, or electrons**.

$$\text{Equivalent weight} = \frac{\text{Molar mass (g/mol)}}{\text{Number of equivalents}}$$

The number of equivalents depends on how many **reactive units** the compound can donate or accept, like **H^+ in acids, OH^- in bases**, or **electrons in redox reactions**.

How do you find that number? Look at the chemical formula:

- For **acids**, count how many **H^+ ions** can be donated.
  → e.g., **H_2SO_4** can donate 2 H^+ → **2 equivalents**

- For **bases**, count the **OH^- ions** it provides.
  → e.g., **$Ca(OH)_2$** gives 2 OH^- → **2 equivalents**

- For **redox reactions**, find how many **electrons** are transferred per mole of reactant.

Why this matters!

Basically, **EW** is used for titrations, redox, and acid-base reactions when it's not just about how many moles you have but about **how many reactive units** are actually involved.

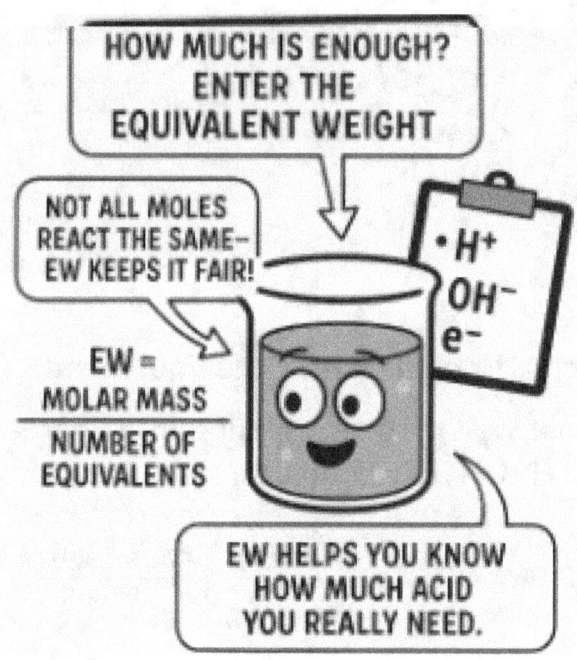

6. Percent Solutions: From Saline to Skincare

Percent solutions are all about ratios. Once you know what is in 100 units of solution, whether grams (**g**) or milliliters (**mL**) you can scale up or down for any recipe, experiment or formulation.

Quick Guide: Which Percent to Use, and When

Type	Use	Example	Formula
w/w% weight/weight	Creams, dry mixes, or DIY cosmetics	Mixing shea butter and zinc oxide to make sunscreen.	$= \left(\dfrac{\text{grams of solute}}{100\,\text{grams of solution}} \right) \times 100$
w/v% weight/volume	Solutions like saline, lab buffers or disinfectants	5% NaCl=5g of salt in 100mL of distilled water	$= \left(\dfrac{\text{grams of solute}}{100\,\text{mL solution}} \right) \times 100$
v/v% volume/volume	Liquid-liquid mixtures like perfumes, essential oils, or alcohol based tinctures	70% ethanol = 70mL ethanol + 30mL water to make 100ml of solution.	$= \left(\dfrac{\text{mL of solute}}{100\,\text{mL of solution}} \right) \times 100$
mg% milligram percent	Medicine and biochemistry, especially for body fluids	Blood glucose = 90 mg% = 90 mg glucose per 100ml of blood	$= \left(\dfrac{\text{mg of solute}}{100\,\text{mL of solution}} \right)$

Bottom Line: Percent solutions are practical, flexible, and everywhere from IV bags to face serums. Once you master how to work with 100 units, you can scale up with confidence.

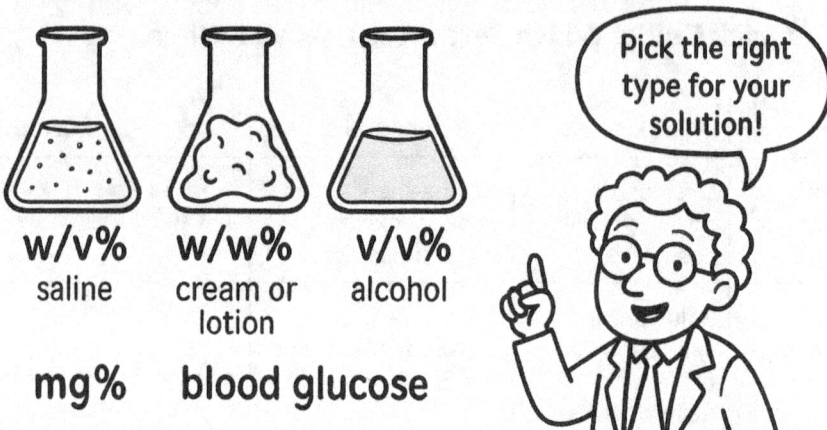

PERCENT SOLUTIONS:
MEASURING WHAT'S IN 100 UNITS

w/v% saline

w/w% cream or lotion

v/v% alcohol

mg% blood glucose

Pick the right type for your solution!

Lab Tip: Know your solution type: **w/v**, **v/v**, or **w/w**, before scaling up any recipe. (*Flip to Quick Reference Guide.*)

7. Osmolarity: The Particle Counter That Keeps Cells Alive

Meet **Osmolarity** the unsung hero quietly protecting your cells from bursting or drying out. If it touches your cells, Osmolarity is on the case. This isn't just chemistry, **it's molecular crowd control**. Think of it as your solution's superpower: **keeping water where it belongs.**

When a solution enters the body, Osmolarity runs the show, counting how many particles are in every drop. It doesn't care what the label says it wants to know how many actual **ions or molecules** are floating around.

Osmolarity = Molarity X Number of particles per formula unit

Examples in Action:
- **1 M NaCl** splits into Na^+ and Cl^- → **2 Osm**
- **1 M glucose** stays whole → **1 Osm**

What's at Stake if Osmolarity is not optimized?
- **Too many particles?** Water rushes out of cells → (hypertonic)
- **Too few?** Cells swell and pop → (hypotonic)
- **Just right?** Cells stay happy, balanced, & thriving → (isotonic)

That's **osmotic pressure,**

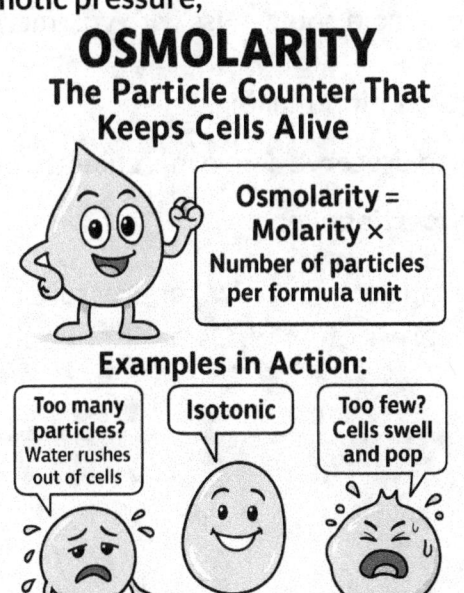

OSMOLARITY
The Particle Counter That Keeps Cells Alive

Osmolarity = Molarity × Number of particles per formula unit

Examples in Action:

Too many particles? Water rushes out of cells

Isotonic

Too few? Cells swell and pop

and **osmolarity** is the gatekeeper.

Where Osmolarity Fights for Balance:

IV fluids – saving red blood cells from rupture
Eye drops – soothing without stinging
Skincare serums – hydrating, not dehydrating
Lab buffers – protecting experimental cells
Electrolyte drinks – restoring fluid balance after workouts

8. Molality: When You Can't Trust Volume

Sometimes, volume lies especially when things get hot (or cold). That's why we have **molality**, the concentration unit that doesn't flinch when temperatures shift.

Molality (m) is the number of **moles of solute per kilogram of solvent** (<u>NOT SOLUTION!</u>). It's especially useful in experiments involving **temperature changes**, where liquid volume can expand or contract but mass stays constant.

$$\text{Molality (m)} = \frac{\text{moles of solute}}{\text{kilograms of solvent}}$$

<u>Why Use Molality?</u> **Molality** is temperature-independent. That makes it ideal for situations where heat or cold could mess with volume, like:

1. **Heating emulsions** during lotion-making.

2. **Cooling gels or serums** to preserve active ingredients.

3. Studying **colligative properties** like:

   a. Boiling point elevation   c. Vapor pressure lowering
   b. Freezing point depression   d. Osmotic pressure

Skincare Science - Why It Matters : *In skincare*, molality comes in handy when:

1. You're heating or cooling during formulation
2. You want accurate ratios of actives like niacinamide or glycolic acid
3. You're working with thick emulsions where volume measurements can be misleading
4. You want to understand how **osmotic pressure** affects water movement across the skin.

FINAL TAKEAWAY: Even if you don't use molality everyday, knowing it gives you the mindset of a formulator,not just a mixer.

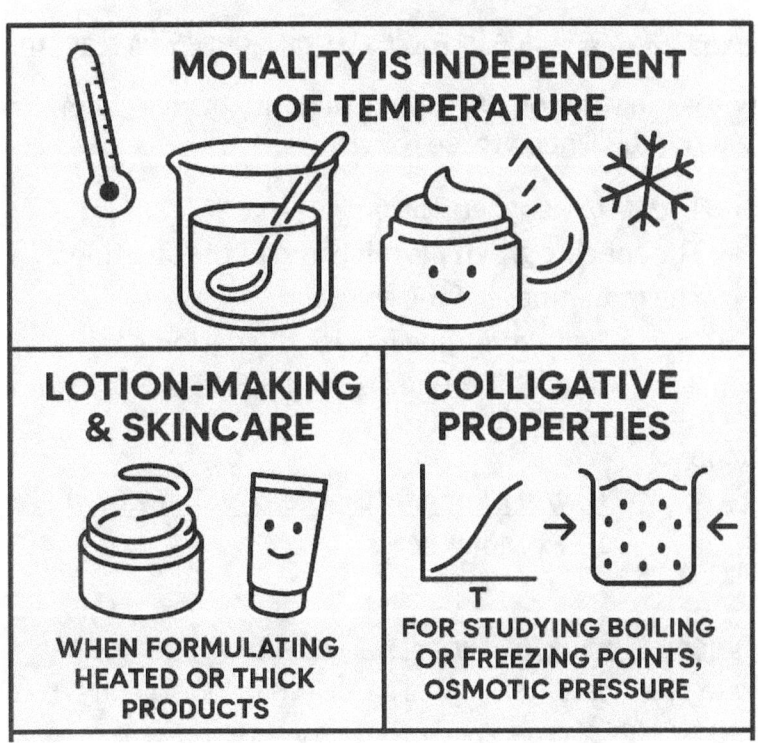

WHY USE MOLALITY?

MOLALITY IS INDEPENDENT OF TEMPERATURE

LOTION-MAKING & SKINCARE

WHEN FORMULATING HEATED OR THICK PRODUCTS

COLLIGATIVE PROPERTIES

FOR STUDYING BOILING OR FREEZING POINTS, OSMOTIC PRESSURE

9. When Weight Matters: %w/w Acids, Density, and Mole Fractions

Earlier, we touched on **%w/w (weight/weight percent)** when talking about creams, dry mixes, and DIY skincare. But now we're leveling up because many **commercial acids** (like HCl, H_2SO_4, and HNO_3) list their concentrations in **%w/w**, not molarity.

To figure out how much to use whether you're preparing a buffer or formulating a serum you'll need to convert between **mass** and **volume**. That's where **density**, or **specific gravity**, comes in.

Key terms to know

ρ (rho): **density** = weight per unit volume
SG (specific gravity) = density relative to water
Since the density of water is **1g/mL**, a solution's **specific gravity is numerically equal to density in g/mL.**

How to Convert %w/w Acids to Moles Using Density (*you got this!*)

To prepare a solution of a known molarity from a commercial acid labeled in **%w/w**, you'll need three things:

- The **%w/w concentration**
- The **specific gravity (or density)** of the solution
- The **molar mass** of the solute
  - *How do you find it?* Use the **periodic table:**
    (e.g., **HCl** = H (1.01) + Cl (35.45) = **36.46 g/mol**)

Formula:
$$Moles = \frac{\% \, w/w \times Specific \, Gravity \times Volume \, (mL)}{100 \times Molar \, Mass}$$

Sample Problem: Making 2 Liters of 0.4 M HCl

How to prepare 2 liters of 0.4 M HCl solution starting from concentrated HCl (28% w/w, specific gravity = 1.15).

Step 1: What Do We Know?

Desired final volume = **2,000 mL**
Desired molarity = **0.4 mol/L**
Starting solution = **28% w/w HCl**
Specific gravity = **1.15 g/mL**
Molar mass of HCl = **36.46 g/mole**

Step 2: Calculate How Many Moles You Need

Moles required = 0.4 mol/L x 2L = 0.8moles

Step 3: Use the %w/w Formula to Solve for Volume

You'll use:
Moles = (%w/w × SG × Volume (mL)) / (100 × Molar Mass)

Solve for Volume:

Volume (mL) = (Moles × 100 × Molar Mass) / (%w/w × SG)

Step 4: Plug in the Numbers

Volume (mL) = (0.8 mol × 100 × 36.46 g/mol) / (28 g/100g × 1.15 g/mL)
= (2916.8 g) / (32.2 g/mL)
≈ 90.6 mL

Final Answer:

To prepare **2 liters of 0.4 M HCl**, measure out **approximately 90.6 mL of 28% w/w HCl** & dilute it with distilled water to a final volume of **2,000 mL**.

** *Lab Safety Tip:* Always add **acid to water**, not water to acid, to avoid splashing or heat-related accidents.

10. Dilutions—Making What You Need from What You've Got

Whether you're in a lab prepping a buffer or in your kitchen formulating a toner, you won't always find the perfect concentration ready to go. That's where the **dilution formula** comes in. From skincare serums to acid solutions, this equation helps you **safely and precisely "water things down"** without losing accuracy:

Formula	You can use this for:
$C_1V_1 = C_2V_2$	* **Molarity (M)** in the lab.
	* **% acids** or **actives** in skincare.
	* mg/mL or w/v concentrations in home recipes.

Just make sure all your units match!

Lab Example: Diluting Sulfuric Acid for a Reaction

How do you prepare **250 mL of 0.2 M H_2SO_4** from a **2 M stock solution**?

Step 1: What Do We Know?

C_1 = 2.0 M (stock concentration) (*aka* 2.0 moles/L)
C_2 = 0.2 M (desired concentration)
V_2 = 250 mL (final total volume)
V_1 = ? (how much stock to use)

Step 2: Use dilution formula

$$C_1V_1 = C_2V_2$$

Step 3: Plug in your numbers

$$(2.0M)(V_1) = (0.2M)(250mL) \Rightarrow V_1 = 50moles/2M = 25mL$$

Step 4 : Measure and dilute

Measure 25 mL of 2.0 M sulfuric acid. Add distilled water to bring total volume to 250 mL

Using Dilution at Home : Same math — different setting!

DIY Example: Diluting a 10% Glycolic Acid Serum to 5%

You bought a **10% glycolic acid** serum, but want a **gentler 5%** version for your skin. The same dilution formula works here.

Step 1: What Do We Know?

C_1 = 10% glycolic acid
C_2 = 5% glycolic acid (your target)
V_2 = 20 mL (how much you want to make)
V_1 = ? (how much of the 10% serum you need)

Step 2: Use the dilution formula:

$C_1 V_1 = C_2 V_2$
$(10)(V_1) = (5)(20) \rightarrow V_1 = 100 / 10 = 10$ mL

Step 3: Make It

- Measure 10 mL of the original 10% serum
- Add 10 mL of distilled water or a neutral base (like aloe or hyaluronic gel)
- Mix gently —> you now have 20 mL of a 5% glycolic acid serum!

Lab Safety Tip : Always **add acid to water, not water to acid, to avoid splashing or heat buildup. (*See quick reference or revisit Chapter 10 for step-by-step.*)

But Wait... Why Does This Work with mL? Isn't M= moles/L?

Great question!

Molarity is defined as **mol/L**, so it's fair to wonder: *Why can I use milliliters here?*

Because you're solving for equal moles on both sides, the **volume units cancel out as long as both V_1 and V_2 are in the same units.**

Mini-Map
Acids & Bases

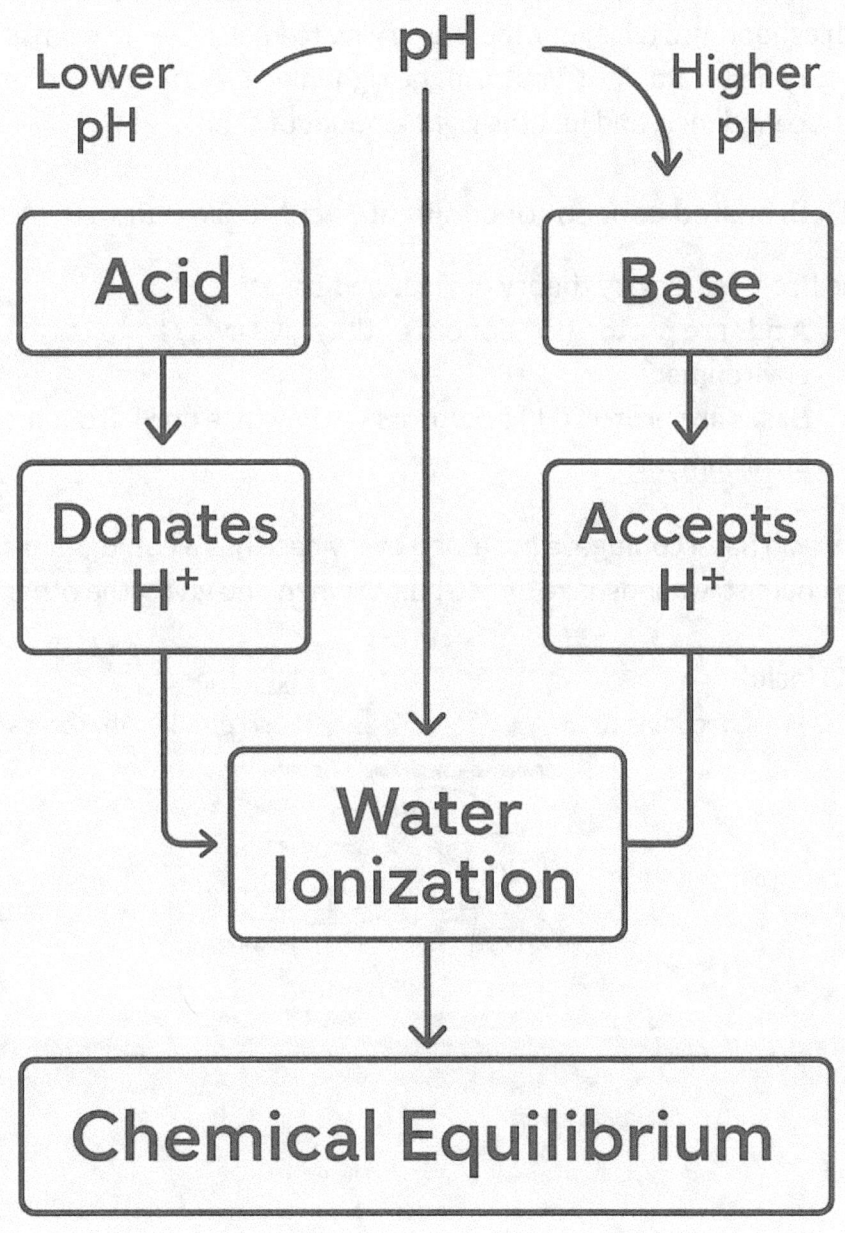

pH

Lower pH

Higher pH

Acid

Base

Donates H$^+$

Accepts H$^+$

Water Ionization

Chemical Equilibrium

Chapter 2 : pH Plot Thickens – Acids, Bases & Buffer Battles

A. The Acid-Base Showdown

Welcome to the ultimate showdown: where acids drop protons, bases catch them, and buffers keep the peace. This chapter takes you through the chemistry of balance, reactivity, and stability. Whether you're troubleshooting a lab experiment or trying to keep a serum from going kablooey and ¡cataplún!, you're in the right place. We've got your back with clarity, confidence, and just the right amount of chaos.

1. Bronsted concept of conjugate acid–base pairs

In the Bronsted-Lowry theory of acids and bases:
- **Acids are proton (H^+) donors** → they give away H^+ to the environment
- **Bases are proton (H^+) acceptors** → they take up H^+ from the environment

Every acid has a **conjugate base**, and every base has a **conjugate acid**. It's like a chemistry handshake or fist-bump: when one gives, the other receives.

HCl (acid)

donates H^+ → becomes Cl^-

NH₃ (base)

accepts H^+→becomes NH_4^+

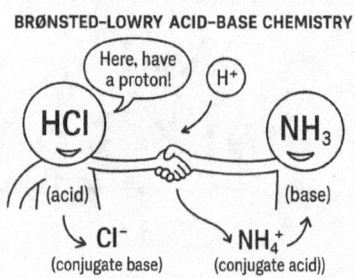

BRØNSTED–LOWRY ACID–BASE CHEMISTRY

FUN FACT: This *donor–acceptor relationship* is what powers acid-base chemistry in buffers, in cells, even in your own stomach acid to the ingredients in your skincare routine.

Fun Fact: This is the same concept that keeps your facial toner pH-balanced or explains why vinegar can neutralize lye in soap-making.

2. Strong vs. weak acids and bases

Not all acids flex the same energy. Some completely dissociate in water, releasing all their H^+ ions with confidence. Others? They hold back, only letting a few protons into the mix.

a. **Strong Acids:** These acids go all in! Every molecule donates a proton (H^+), leaving no original acid behind. Just free ions doing their thing.

Strong Acid Dissociation Reactions

Acid	Dissociation Reaction
HCl (hydrochloric acid)	$HCl + H_2O \rightarrow H_3O^+ + Cl^-$
HNO₃ (nitric acid)	$HNO_3 + H_2O \rightarrow H_3O^+ + NO^-$
HBr (hydrobromic acid)	$HBr + H_2O \rightarrow H_3O^+ + Br^-$
HI (hydroiodic acid)	$HI + H_2O \rightarrow H_3O^+ + I^-$
HClO₄ (perchloric acid)	$HClO_4 + H_2O \rightarrow H_3O^+ + ClO_4^-$

*Strong acids? Total extroverts.** They give away their H^+ the instant they show up to the party.

b. **Weak Acids:** These acids only partially dissociate in water. That means you'll have a mix: some molecules gave up their H^+, others are holding on.

Weak Acid Dissociation Reactions

Acid	Dissociation Reaction
CH₃COOH (acetic acid, aka vinegar)	$CH_3COOH \leftrightharpoons H^+ + CH_3COO^-$
H_2CO_3 (carbonic acid)	$H_2CO_3 \leftrightharpoons H^+ + HCO_3^-$
HF (hydrofluoric acid)	$HF \leftrightharpoons H^+ + F^-$
HCN (hydrocyanic acid)	$HCN \leftrightharpoons H^+ + CN^-$

*Weak acids** are **more introverted. Shy donors.** They only part with a few H^+ ions, and only when nudged.

What This Means in Real Life:

In DIY skincare, **citric acid** (a weak acid) is used to balance pH gently without over-acidifying. While in the lab, you **choose strong acids for full reaction** and **weak acids for controlled conditions**, like in buffer solutions.

* **Important tip:** Strength has **nothing to do with concentration.** A strong acid can be **dilute**, and a weak acid can be **concentrated. It's** all about how much they ionize in water.

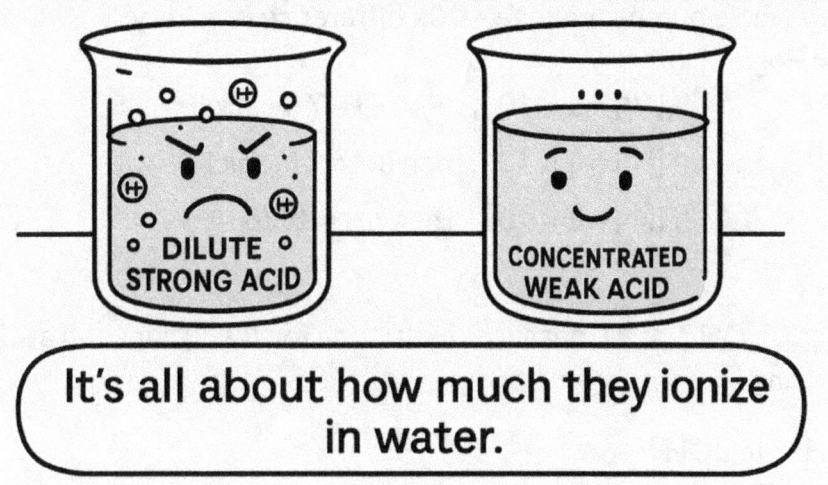

'IMPORTANT TIP: STRENGTH HAS NOTHING TO DO WITH CONCENTRATION

DILUTE STRONG ACID

CONCENTRATED WEAK ACID

It's all about how much they ionize in water.

3. pH and pOH : The Power (*or Lack*) of Hydrogen

Protons (aka **H⁺**), **control** the **pH**, the scale that tells you how acidic or basic your solution is. Whether you're balancing a facial toner, troubleshooting a buffer, or measuring the sting in vinegar, pH is your go-to number.

<u>What is pH ?</u> **pH** stands for "**potential of hydrogen**" and tells you the concentration of H^+ ions in a solution. It's a *logarithmic scale*, which means every whole number change = **10x difference** in acidity:
pH = -log[H⁺].

If $[H^+] = 1 \times 10^{-7}$, then **pH = 7** (*neutral*)

If $[H^+] = 1 \times 10^{-3}$, then **pH = 3** (*acidic*)

If $[H^+] = 1 \times 10^{-11}$, then **pH = 11** (*basic*)

<u>What is pOH ?</u> **pOH** measures **hydroxide ion (OH⁻)** concentration, *yin to pH's yang.*

pOH = -log[OH⁻]

They are connected through this nifty formula: **pH + pOH =14**

$$pH = -\log[H^+]$$

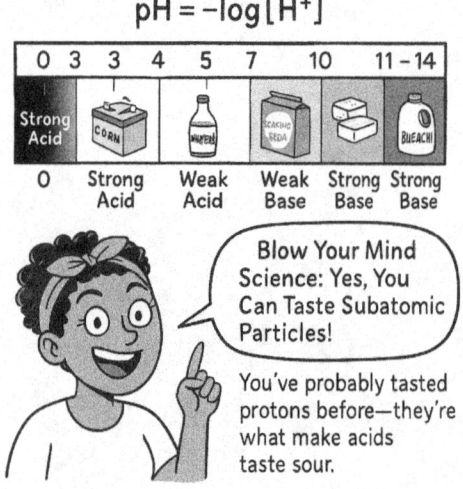

| 0 | 3 | 3 | 4 | 5 | 7 | 10 | 11 – 14 |

Strong Acid

| 0 | Strong Acid | Weak Acid | Weak Base | Strong Base | Strong Base |

Blow Your Mind Science: Yes, You Can Taste Subatomic Particles!

You've probably tasted protons before—they're what make acids taste sour.

Blow Your Mind Science: Flavor Chemistry Fun Fact!

Sour taste in lemon? That's your tongue detecting acidity, aka a high *concentration of H⁺ ions*.

Bitter or soapy taste? That's often a base, like lye, forming soaps with fats.

Lab Hack: pH meters or pH paper (*litmus or universal indicators*) give you a fast read on acidity. For accurate measurements, use a calibrated meter.

DIY Realness:

Formulating skincare? Haircare? Cleaning solutions? **pH matters**.

- Acidic products (like *Vitamin C serums*) help exfoliate.
- Base, *aka alkaline*, like soaps can strip oils. **Stay between 4.5 and 6.5 to stay skin-friendly.**

4. Ionization of water: When H_2O Splits the Bill

Water (aka *aguacita*), isn't just the **universal solvent**, it's also a *low-key superhero* that **self-ionizes**. In every drop of pure water, a tiny fraction of **H_2O molecules split into two ions:**

$$H_2O \rightleftharpoons H^+ + OH^-$$

Technically, the **H^+** quickly bonds to another **H_2O** to form **H_3O^+** (hydronium), but we often shorthand it as **H^+** to keep the math simple.

The "WHY" water is neutral

This **self-ionization** is what gives water a **neutral pH of 7.** In **pure water, at 25°C,** water **dissociates** to $[H^+] = [OH^-] = 1 \times 10^{-7}$ mol/L.

Multiply the two together: $[H+] \times [OH^-] = 1 \times 10^{-14}$. This value is called **Kw,** the **ion product constant for water** (*valid ONLY at 25°C*).

Why You Should Care (*Even If You're DIYing*):

- This is the **math behind neutral pH.**
- It's why **water can shift pH** slightly when ingredients are added.
- It powers reactions in **buffers, serums, shampoos,** and even your own **blood.**

Your face mist or toner might look like "just water,", but even a drop of essential oil or acid changes the pH because water is always in flux.

Fun Fact: Your face mist or toner might look like "*just water*," but even a drop of essential oil or acid changes the pH because water is always in flux.

Always remember!

Even "pure" water has ions floating around.Thanks to auto-ionization, **pH + pOH = 14** holds true for all aqueous solutions at 25°C.

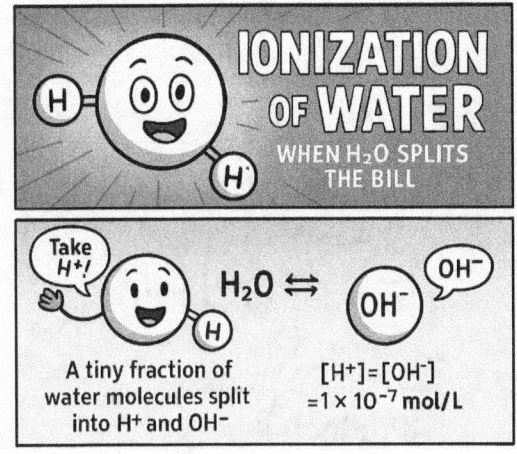 

B. Plot Twist: Equilibrium Enters

Just when you thought acids and bases were done with their drama, equilibrium steps into the spotlight. This is where chemistry shows off its balancing act because not all reactions go to completion. Many reach a point where the forward and reverse reactions happen at the same rate. That point? It's called **equilibrium.**

1. Equilibrium Constants (Keq)

At equilibrium, the concentrations of reactants and products stay constant, not because the reaction stops, but because it's happening in both directions equally.

For a _general reaction_: $aA + bB \rightleftharpoons cC + dD$
The **equilibrium constant, Keq,** is calculated as: $K_{eq} = ([C]^c [D]^d) / ([A]^a [B]^b)$.
The brackets [ ] mean "**concentration**," or **how much of each substance is present in the solution.**

So basically, a **reversible reaction** is like a tug-of-war.

- If **Keq > 1**, the products pull harder and win
- If **Keq < 1**, the reactants hold their ground.

Why It Matters to Makers:

Keq tells you if your formula stays stable or if it might shift over time. Understanding equilibrium helps when you're blending ingredients that might react, like acids with bases in toners, serums, or natural cleaners. It's the quiet chemistry that decides if your product remains balanced... or *fizzles out.*

2. Acid & Base Equilibria: How Ka and Kb Measure Strength (Without the Drama)

Not all acids go full drama mode, especially the weak ones. Weak acids and bases don't completely dissociate in water. Some molecules break apart into ions, others hang back and stay whole. That's where equilibrium comes in. To measure how much they dissociate, we use special equilibrium constants:

Ka = acid dissociation constant;
= how much an acid dissociates into H^+

Kb = base dissociation constant
= how much a base dissociates into OH^-

Bigger values = more dissociation = stronger acid/base

These values help us quantify the strength of weak acids and bases.

Substance	Type	Ka or Kb	Dissociation
HCl	Strong Acid	Very large (>>1)	Complete
HNO_3	Strong Acid	Very large (>>1)	Complete
CH_3COOH	Weak Acid	1.8×10^{-5}	Partial
HF	Weak Acid	6.6×10^{-4}	Partial
NaOH	Strong Base	Very large (Kb)	Complete
NH_3	Weak Base	1.8×10^{-5} (Kb)	Partial

Let's compare with examples:

Type	Example	Dissociation Reaction	Ka Explanation
Weak Acid (used in kitchen or cosmetics)	Acetic Acid (vinegar)	$CH_3COOH \rightleftharpoons CH_3COO^- + H^+$	**Ka** = [CH_3COO^-] [H^+] / [CH_3COOH] **Ka = 1.8 × 10^{-5}** (Double arrow = reversible)
Strong Acid (lab only, NOT for skincare)	Hydrochloric Acid (HCl)	$HCl \rightarrow H^+ + Cl^-$	**No Ka** needed; dissociates completely in water. (Single arrow = complete dissociation in water)

Ka is a constant that tells how much a weak acid dissociates. For example, acetic acid has Ka ≈ 1.8 × 10^{-5}. Strong acids don't need a Ka because they fully dissociate.

Why DIYers and Makers Should Care:

- **Skincare:** Acids like glycolic or lactic are weak acids. Their **Ka values** affect how deeply they exfoliate your skin.

- **Cleaning:** Understanding **Kb** helps you pick the right base for cutting grease without being too harsh.

- **Serums & Solutions:** Knowing **Ka** and **Kb** helps prevent unwanted pH shifts or surprise reactions when you mix ingredients.

VERY IMPORTANT TO REMEMBER: If an acid has a huge **Ka** (or no listed **Ka** because it dissociates fully), it's very strong think **HCl** (hydrochloric acid), **HNO₃** (nitric acid), **H₂SO₄** (sulfuric acid) (first proton only). **These are lab tools, not cosmetic or food ingredients!**

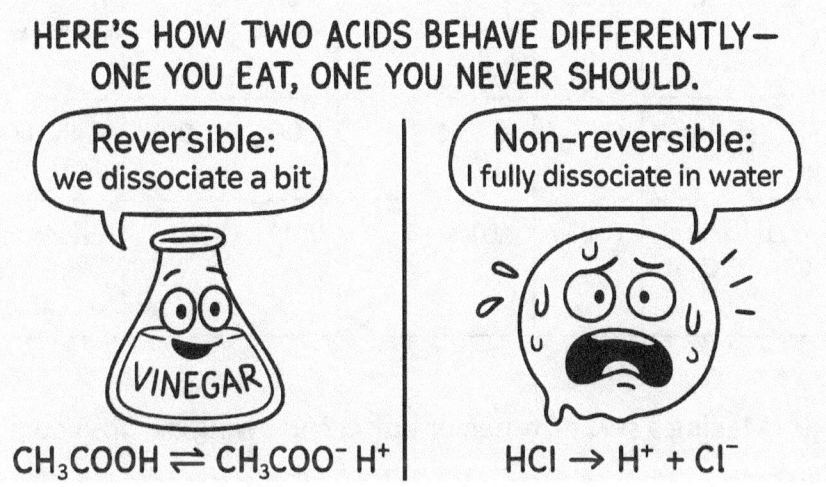

3. Degree of Dissociation (α): How Much Actually Happens

Not all acids or bases fully break apart in water. That's where the **degree of dissociation (α)** comes in. It tells you what fraction of the acid or base actually ionizes, aka how much is doing the chemistry work.

Formula: α = (amount dissociated) / (initial concentration)

How to interpret α:

If $\alpha \approx 1$, the substance fully dissociates → it's strong.
If α is small (e.g. 0.01 or 0.1), only a little dissociates → it's weak.

Real World Examples:

Scenario	Acid	[ ]	Amount Dissociated	α	Meaning
DIY Skincare Toner	Lactic acid (AHA)	0.1 M	0.006 M	**0.06** → 6%	Mild exfoliation, gentle on skin
Lab Solution	HCl (strong acid)	0.1 M	0.1 M	**1.00** → 100%	Fully dissociates, very acidic
Kitchen Vinegar	Acetic acid (CH_3COOH)	0.1 M	0.004 M	**0.04** → 4%	Weak acid, food-safe and versatile

**DIY Insight:** Making a serum, toner, or buffer? Knowing **α** helps you predict pH shifts.

High **α** = big pH change

Low **α** = gentle adjustment

This helps you **design formulas** that are effective **without overdoing it,** whether you're in the lab, the kitchen, or your skincare corner.

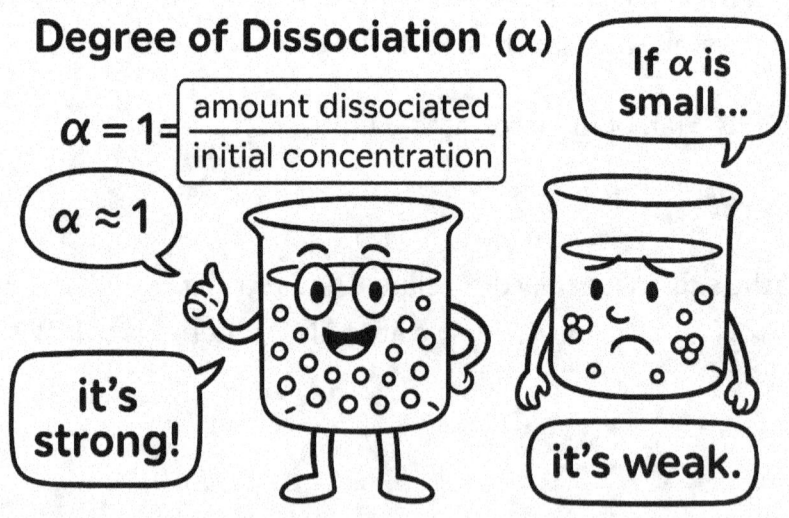

4. Hydrolysis of Salts: Water Gets Involved

When certain salts dissolve in water, their ions may react with H_2O in a sneaky way shifting the pH. This process is called **hydrolysis**.

1) _Example (from a weak acid):_

$$CH_3COONa \rightarrow CH_3COO^- + Na^+$$
$$CH_3COO^- + H_2O \rightleftharpoons CH_3COOH + OH^-$$

Result: The solution becomes basic because **OH$^-$** is produced.

2) _Now flip it: (from a weak base):_

$$NH_4Cl \rightarrow NH_4^+ + Cl^-$$
$$NH_4^+ + H_2O \rightleftharpoons NH_3 + H_3O^+$$

Result: The solution becomes acidic because H_3O^+ is produced.

Why It Matters

Hydrolysis is **why:**

- Baking soda (a weak base salt) can spike pH in DIY skincare.
- Some "natural" deodorants cause irritation; they may shift pH too far.

- Buffer systems work to resist wild pH swings by balancing these hydrolytic shifts.

Salt Hydrolysis: Which Salts Shift pH (& *How*)

#	Salt Type	Example	Ion Hydrolyzed	Reaction with Water	pH
1	From Strong Acid + Strong Base	Sodium Chloride (NaCl)	None	No reaction, remains neutral	Neutral (=7)
2	From Weak Acid + Strong Base	Sodium Acetate (CH₃COONa)	CH_3COO^-	$CH_3COO^- + H_2O \rightleftharpoons CH_3COOH + OH^-$	Basic (>7)
3	From Strong Acid + Weak Base	Ammonium Chloride (NH₄Cl)	NH_4^+	$NH_4^+ + H_2O \rightleftharpoons NH_3 + H_3O^+$	Acidic (<7)

DIY Insight: *Making face masks, toners, or soaps*? If your formula includes citric acid, ammonium salts, or baking soda, you're already in the world of hydrolysis. Knowing what your ingredients do in water helps you avoid surprise pH swings.

Fun Chemistry Fact: Hydrolysis is where equilibrium and dissociation meet water. It's like a remix of all the chemistry concepts we've just covered: **Ka, Kb, equilibrium,** and **ionization**. That's why it's the foundation of buffer solutions.

Fun Chemistry Fact:
Hydrolysis is where equilibrium and dissociation meet water.

Mini-Map

BUFFER LEAGUE – SCIENCE'S PEACEKEEPPERS

WHAT IS A BUFFER (AND HOW IT WORKS)

HENDERSON-HASSELBALCH EQUATION

$$pH = pk + \frac{[A-]}{[HA]}$$

TITRATION OF A WEAK ACID

BUFFER CAPACITY & REAL-WORLD EXAMPLES

C. Buffer League – Science's Peacekeepers

1. What is a Buffer and How It Works

Imagine a pH bodyguard. That's what a buffer is!
A buffer solution keeps the peace resisting wild pH swings when small amounts of acid or base show up.

The secret?

A dynamic duo: A weak acid and its conjugate base OR A weak base and its conjugate acid. Together, they step in to **neutralize** added acid (H^+) or base (OH^-) and keep the solution's pH stable.

Think of **buffers** like chemistry's emotional support system. When the lab gets stressful, they stay calm and keep pH steady.

Common Buffer Pairs:
- Acetic acid (CH_3COOH) and sodium acetate (CH_3COONa)
- Ammonia (NH_3) and ammonium chloride (NH_4Cl)

How it works:
- Add an acid (H^+)? → The base part of the buffer soaks it up.
- Add a base (OH^-)? → The acid part steps in to neutralize it.

Why Buffers Matter:

Keep enzymes from denaturing   Stabilize skincare products

Maintain your blood at ~pH 7.4   Protect fragile reactions in the lab

Take home message : Buffers => Chemistry's Peacekeepers. When pH chaos threatens, this duo saves the day.

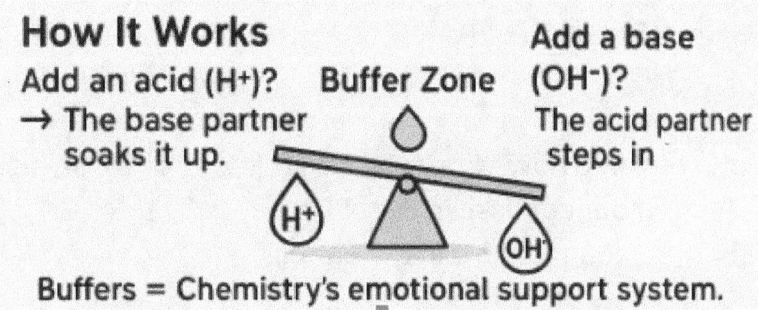

How It Works

Add an acid (H+)?
→ The base partner
  soaks it up.

Buffer Zone

Add a base (OH⁻)?
The acid partner
steps in

Buffers = Chemistry's emotional support system.

∿ Citizen Science Spotlight: Soil Buffers in Action

Soil isn't just dirt…it's chemistry in motion. Healthy soil contains natural buffers like carbonates, clay, and organic matter that **resist wild pH swings**, just like the buffer you build in the lab or in your kitchen

Example from the Field: During my Arctic citizen science research, I tested soil pH before and after adding a weak acid. The pH barely moved. *Why?* Because the soil's natural buffer absorbed the acid just like your buffer duo would.

Try it yourself:
1. Take a soil sample from your yard or garden.
2. Mix it with distilled water and test the pH.
3. Add a few drops of diluted vinegar. Retest.
   Did the pH change much?
       If not, you've got buffering power beneath your feet!

Want to contribute real data to a global soil project? Visit soilsciencelab.com to join our citizen science network!

2. Henderson-Hasselbalch Equation-The Buffer Balancer

When you want to hold pH steady, this is your go-to formula for building or adjusting a buffer:

$$pH = pKa + \log([A^-]/[HA])$$

Where:
- **pKa** tells you how strong the acid is (lower pKa = stronger acid)
- $[A^-]$ = amount of **base** in the buffer
- $[HA]$ = amount of **acid** in the buffer

Picture a seesaw between acid and base.

This formula helps you keep it perfectly balanced so your pH doesn't tip too far in either direction.

When Is This Formula Useful?

Use *Henderson-Hasselbalch* when you want to:
- Make a facial toner that stays gentle (even over time).
- Prep an enzyme bath that only works at a specific pH.
- Keep your kombucha acidic enough to be safe but not mouth-puckering.

Lab Example: Designing a Buffer at pH 5.0 Using Acetic Acid

How do you build a buffer that holds at pH 5.0 using acetic acid (pKa ≈ 4.76)?

Step 1: What Do We Know?

Target pH = 5.0
pKa of acetic acid = 4.76
We're solving for the **ratio of base to acid** $\rightarrow [A^-]/[HA]$.
(This tells us how much **conjugate base** we need compared to **acid**.)

Step 2: Use the Henderson-Hasselbalch Equation:

$$pH = pKa + \log([A^-]/[HA])$$

Step 3: Plug In Your Numbers

$5.0 = 4.76 + \log([A^-]/[HA])$ → *Subtract* **4.76** from both sides:

$0.24 = \log([A^-]/[HA])$ → Raise both sides as powers of 10:

$[A^-]/[HA] = 10^{0.24} \approx 1.74$

Step 4: What It Means

You need your **conjugate base** concentration to be about **1.74 times greater** than your acid.

If you use:

> 1 mole of acetic acid → use **1.74 moles** of sodium acetate.
> 10 mL of acid → use **17.4 mL** of base solution (*adjusted for molarity*).

Lab Tip: Use this ratio when mixing buffer components to hit the exact pH you need. Always double-check the pKa and use fresh solutions for accuracy **especially for enzyme work.**

Why It Matters:
- Helps your products stay safe and effective.
- Keeps your experiments predictable and reliable.
- Prevents surprise pH swings after storage, heat, or time.

Take-Home Message:
Henderson-Hasselbalch ⇒ Use it when you want your solution to stay balanced even when life throws acid or base at it.

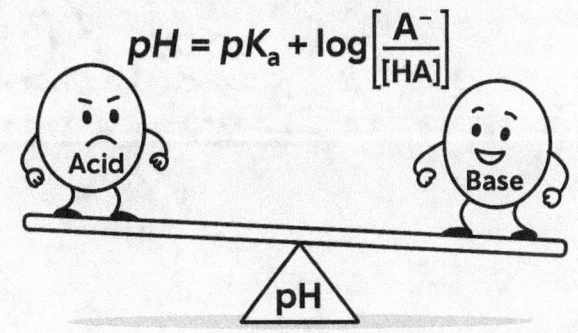

YOUR pH PEACEKEEPING EQUATION

Use it when you want your solution to stay balanced even when life throws acid or base at it.

$$pH = pK_a + \log\left[\frac{A^-}{[HA]}\right]$$

Acid

Base

pH

3. Titration of a Weak Acid – Finding the Buffer Sweet Spot

Titration is like a slow-motion chemistry dance between acid and base. You add one drop at a time watching how pH changes along the way. For weak acids, the pH change makes a smooth S-shaped curve.

The **magic moment** is the halfway point: **Half-equivalence point**
That's when **[HA] = [A⁻]** and **pH = pKa** —>: *maximum buffering power*.

Real-World Example:
Neutralizing vinegar (acetic acid) with baking soda (NaHCO₃)?

 You'll notice the **pH levels off around 4.7** which just happens to be the **pKa of acetic acid.** That's your **buffer zone**…right where the solution can resist pH swings the best.

Why It Matters:

- Helps you **design buffers** at their strongest point.
- Shows you the **sweet spot** where acid/base are perfectly balanced.
- Lets you **measure pKa** experimentally if you don't know it.

4. Buffer Capacity - How Much Can Your Buffer Take?

So you've built a buffer…great! But here's the next big question, **How much can it actually handle before the pH starts to swing?** That's called **buffer capacity** and it's all about how much your buffer can absorb **before it reaches its limit.**

Think of it like the size of a sponge:

Small sponge?
It gets soaked fast.

Big sponge?.
It holds more.

What Affects Buffer Capacity?

Buffer muscle comes from how much acid and base you've got and how close they are to a 1:1 ratio. More balance => less pH drama.

When Buffers Save the Day

- **In your blood:** The **bicarbonate buffer system** keeps pH steady around 7.4 so your cells stay alive and well.
- **In skincare:** Buffers keep **Vitamin C serums stable, effective, and gentle to your skin.**
- **In the lab:** Buffers protect **delicate enzymes and proteins.**
- **In cleaning:** pH-balanced products protect your **hands, surfaces, and materials.**

Take-Home Message

Buffers aren't jargon, they're **pH lifesavers**. Whether you're in the lab or your bathroom, Knowing **buffer capacity** gives you **superhero-level control.**

Final Word from Aidyl
¡WEPA! You did it!

You just mastered the math that powers everyday science, no gatekeeping, no fluff. You decoded pH, balanced equations, and calculated your way through acids, buffers, and even kombucha.

So now I ask you:

What will you create next? A serum? A startup? A smarter lab? A better world?

Whatever it is, the math's got your back. Because now you **understand** it and **speak its language.** And here's the beautiful part: **Science moves. So do you.**

Want to put your skills into action? I lead a global citizen science project studying soil health, microplastics, and the chemistry of our changing world. You can collect real data and be part of something bigger whether you're in a lab, a classroom, or your own backyard.

Join us at **soilsciencelab.com**

Keep going. You're ready. You learned all this

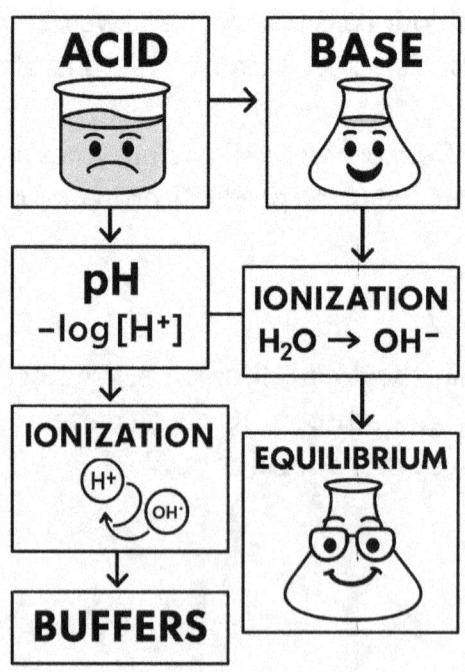

Quick Reference Guide

Because even superheroes need a utility belt.

Use this quick reference guide as your go-to when the math gets real, and the pipettes start flying. These pages recap the formulas, units, and conversions you've seen throughout the book all in one place.

Concept	Formula	Notes
Molarity (M)	M = moles / volume (L)	Most common unit in solution chemistry.
Dilution	$C_1V_1 = C_2V_2$	Keep units consistent.
Molality (m)	m = moles / mass of solvent (kg)	Best for temperature-sensitive work.
Normality (N)	$N = n \times M$	n = number of equivalents
Osmolarity	Osm = M × # of particles	Count dissociated ions!
Activity (a)	$a = \gamma \times [C]$	γ = activity coefficient
Ka / Kb	$Ka = [A^-][H^+] / [HA]$	Measure of acid/base strength.
pH / pOH	$pH = -\log[H^+]$, $pOH = -\log[OH^-]$	pH + pOH = 14 (at 25°C)
Percent Solutions	See cheat sheet below	Know your w/v, v/v, and w.w.

Common Lab Math Formulas

Units & Conversions

Prefix	Symbol	Value	Example
kilo	k	1,000	1 kg = 1,000 g
centi	c	0.01	1 cL = 0.01 L
milli	m	0.001	1 mL = 0.001 L
micro	µ	10^{-6}	1 µg = 0.000001 g
nano	n	10^{-9}	1 nL = 0.000000001 L

Cheat Sheet: % Solutions

Type	Description	Example
w/v%	g solute / 100 mL solution	5% NaCl = 5 g in 100 mL H_2O
w/w%	g solute / 100 g solution	70% glycolic acid in skincare
v/v%	mL solute / 100 mL solution	70% ethanol = 70 mL EtOH + 30 mL water
mg%	mg solute / 100 mL solution	90 mg/dL glucose in blood

Vocabulary Power-Up

Science Terms That Sound Scarier Than They Are.

Use this page like a decoder ring for chemistry conversations. Whether you're DIY-ing a toner or titrating in the lab, these words have your back.

Core Concepts

Aqueous Solution – A mixture where something is dissolved in water.

Solute – The stuff that gets dissolved (like salt or sugar).

Solvent – The liquid doing the dissolving (usually water).

Molarity (M) – Moles of solute per liter of solution.

Molality (m) – Moles of solute per kilogram of solvent.

Normality (N) – Equivalents of reactive units (H^+, OH^-, or electrons) per liter of solution.

Equivalent (eq) – The amount of substance that reacts with 1 mole of H^+, OH^-, or electrons.

Osmolarity – Total particles in a solution that affect water movement (osmosis).

Acid-Base Super Squad

Acid – A proton (H^+) donor.

Base – A proton (H^+) acceptor.

Conjugate Acid/Base – The acid/base partner created after a proton is transferred.

Strong Acid/Base – Fully dissociates in water (no chill).

Weak Acid/Base – Partially dissociates (has boundaries).

pH – "Power of hydrogen"; measures how acidic or basic a solution is.

pOH – Measures hydroxide ion concentration; opposite of pH.

Ka / Kb – Strength of a weak acid (Ka) or base (Kb) in water.

pKa – The pH at which an acid is half dissociated; smaller pKa = stronger acid.

Buffer – A pH bodyguard: resists sudden pH changes using a weak acid/base pair.

Buffer Capacity – How much pH stress a buffer can handle before it gives in.

◎ Other Chemistry MVPs

Dissociation – When a molecule splits into ions in solution.
Equilibrium – When forward and reverse reactions happen at the same rate.
Degree of Dissociation (α) – How much of an acid/base actually ionizes.
Activity (a) – The effective concentration of a substance, accounting for real-world interactions.
Density (ρ) – Mass per unit volume, often needed for converting %w/w to molarity.
Specific Gravity (SG) – Density compared to water (water = 1.00).
Hydrolysis – When water reacts with a salt to shift pH.
Percent Solutions
- **w/v%** = grams per 100 mL
- **v/v%** = mL per 100 mL
- **w/w%** = grams per 100 g

Pro-Tip: If it has a log, a bracket, or a weird Greek letter it's probably just math telling you how much or how strong something is.

You got this!